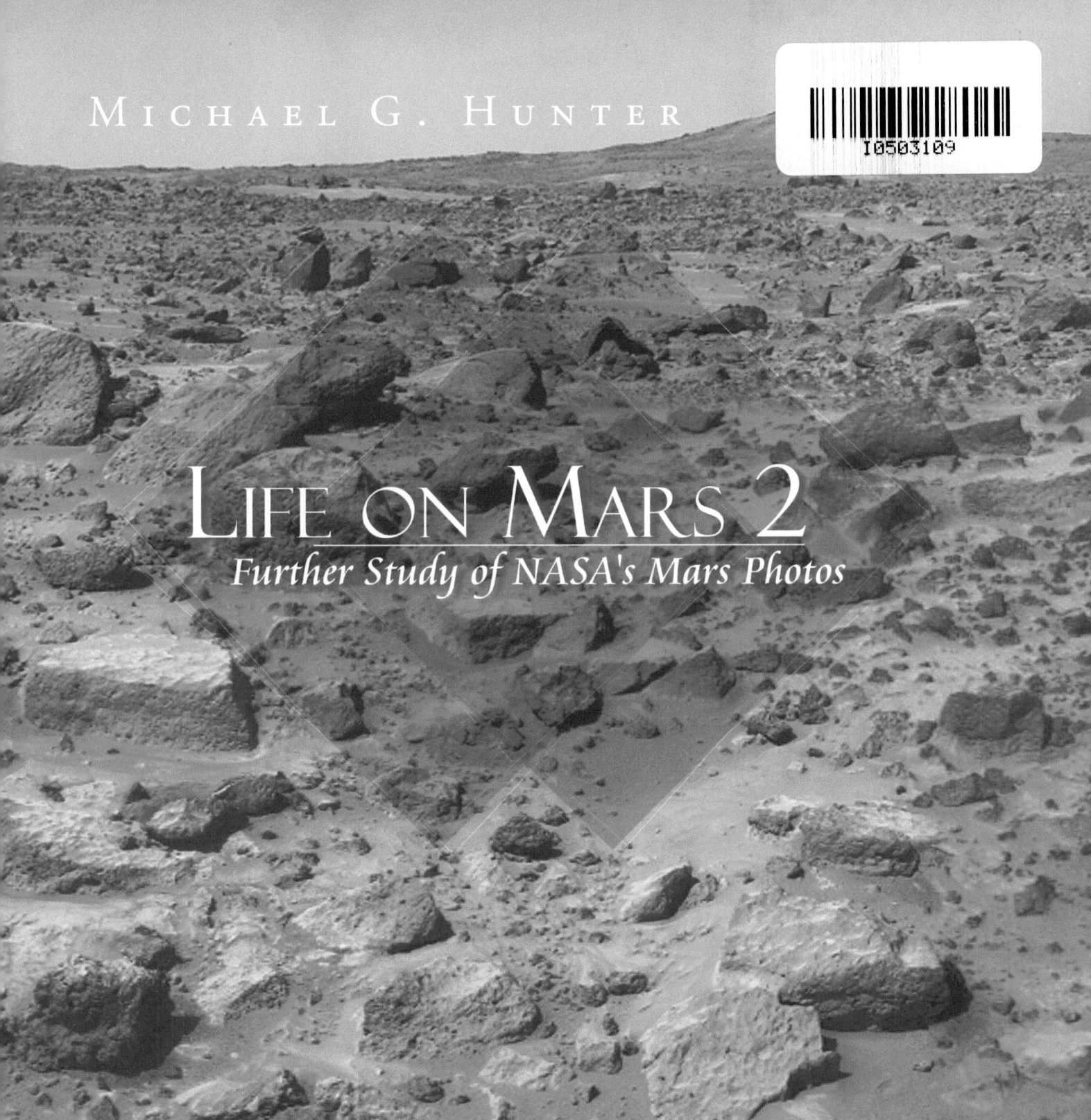

MICHAEL G. HUNTER

LIFE ON MARS 2
Further Study of NASA's Mars Photos

To order additional copies of this book, contact:
Xlibris Corporation
1-888-795-4274
www.Xlibris.com
Orders@Xlibris.com

LIFE ON MARS 2

Michael G. Hunter

INTRODUCTION

On Earth, thiings seem simple enough. Rocks are rocks. Fish swim in the water, animals crawl on the ground, and birds fly in the air. People live in houses, wear more clothing when it's colder, normally grow to five or six feet tall, and avoid direct sunlight because they know solar radiation causes problems like skin cancer. On Mars, things are not so simple. Fish flop around on the surface of the desert, while animals, birds, and even people swim around in the sand. Little things are big and big things are bigger. 5000 year-old cultures still persist. People wear summer clothing in a climate more severe than winter in Antartica. Some people cannot stand up because their heads would be above the atmosphere. Animals extinct on Earth still thrive on Mars. Many animals on Mars look like rocks. Everything is haywire, backwards, upside-down, absurd, wacky, impossible, and bizzare. Life on Mars is a paradigm shift of major proportions which will likely take awhile to register, gain scientific approval, and eventually cause significant repercussions in many areas of science and knowledge.

What does it mean to our everyday-everyone's lives, now that we know for sure that Mars is inhabitable by humans? For the most part, it changes nothing. We still go about our lives, worried about finances, love life, pets, cleaning our homes, mowing our lawns, politics, whatever. But in perspective, think of the similar event in 1492, when the New World, the Western Hemisphere, was "discovered". Of course it had magnanimous repercussions. It opened up new hopes and dreams for people. It led to great expansion and wealth, eventually, considering the bountiful resources that were ultimately found and exploited. People who were suffering in despair, discontent with their livelihoods, gave up everything to go to the new world of opportunity. Is Mars like that? Perhaps. Obviously it's a haven for scientists. But are there resources that can be exploited? Is it a prime vacation hideaway? A source of rare minerals and gemstones? Or, is it a land of horror, death and discomfort, like a living hell? Only time will tell. But knowing people, through history, we can expect a vast new hope is dawning; whether it will be fulfilled or not is the only question.

As I finished that first book, Curiosity was at the site they call Rocknest. The photo that showed the rover scooping soil happened to include a Martian head, a diver, in the sand. I managed to put that photo of a human in my first book. Only after much study did I realize the meaning of that, and only several weeks later did I realize that the man was dead. I subsequently realized that in the photo at Rocknest, there were a few animals and humans that I needed to include in this book 2.

Most recently, the Curiosity rover was photographing areas containing pools of liquid water in Yellowknife Bay, (which NASA fails to mention in its photo captions). As I wrote in "Life on Mars", "...I believe the Curiosity rover will cross paths with waterways, animals, plants and humans, in the very near future..." Also, the rover was studying the atmosphere carefully, analyzing samples of air and finding pretty much what has been found several times before. There is evidence of ancient water in abundance in the southern hemisphere. Carbon dioxide seems pretty plentiful on Mars, with 96 percent of the atmosphere being that gas. Oxygen, on the other hand, is quite rare, forming only .14 percent, less than one percent of the oxygen in our atmosphere on Earth. It would seem that any animal life on Mars would be near suffocation from the lack of oxygen, and under great stress from the low atmospheric pressure. (In the Swartzenegger movie "Total Recall" the main characters were exposed briefly to Mars' raw conditions, with their swollen bodies and bulging eyes!) So of course, with photos showing people going about Mars, similarly to those on Earth, something has been miscalculated and misjudged, and this newfound awareness must be immensely valuable to scientists. But strangely, even at this writing in early 2013, there is still no mention in NASA's writings of existing water or animals on Mars. They are suggesting lately that they may have found microbes in the atmosphere and organic compounds in the soil, (indirect signs of life). Nevertheless, I will continue in my attempt to show what is going on with the living beings on Mars. Perhaps the rover is a stone's throw from the water in Gale Crater, and will photograph the lake any day now with myriads of animals thriving in and around it. In a way, I feel like a reporter in a strange never-before-seen land, or perhaps like James Darwin, discovering new species and trying to study and document them.

Also, one of the photos from the Pathfinder rover, which beamed images from Mars from July 4, 1997, until it's batteries died in September of that year, became the cover for my first book. That photo, chosen by the Xlibris publication artist, as well as other Pathfinder images, led me to more findings regarding life on Mars, which I feel compelled to share in this book. And with findings of animals and humans in the 1997 photos, the question then arose, were there photos earlier than 1997 that showed animals and humans on Mars? So, I investigated some earlier photos and guess what? My findings are in this book.

Every time NASA puts a new photo in their gallery, showing the dry sands and rocks of Mars, it excites me, because I want to see what new or familiar type of animal is roaming there, and what more can be deduced from the images regarding human level of development there. There is, I must admit, a fear that develops in studying a newfound type of animal or people. Would they be dangerous to us if we were there? Are they something we need to be aware of and avoid, or are they something we need to learn about more and eventually greet and communicate with? Do they have our current capabilities of space travel, or are they less advanced technologically than ourselves? I try to answer those questions in this book.

One of the last entries in my first book was about the long-span bridge that the road has as it crosses the waterway. A closer look at that photo reveals what I came to believe is a city. And so, day by day, the discoveries continue long after my first book was in publication, with much more to explain.

Furthermore, I did some research about the Nephilim and the mechanics of Earth/Mars exchanges. My theories hopefully give some direction in the solution of the mystery of the Nephilim, as well as the mystery of how the various Earthlike animals and dinosaurs came to exist on Mars. How did they get there? It does not seem likely that animals could evolve on Mars to become exactly the same as those on Earth.

The findings from Pathfinder images, Viking Lander images, Mariner flyby, and latest Curiosity rover images, new animal images, additional human images, the city, the search at Rocknest, ancient Egyptian evidence, and my theories of the Nephilim and Earth/Mars Exchanges, all came together in this second book. Like "Life on Mars", this book, "Life on Mars 2", is a continuing documentation of discovery of monumental importance to mankind; yet it is no more than a scratch on the surface of our brother planet, Mars.

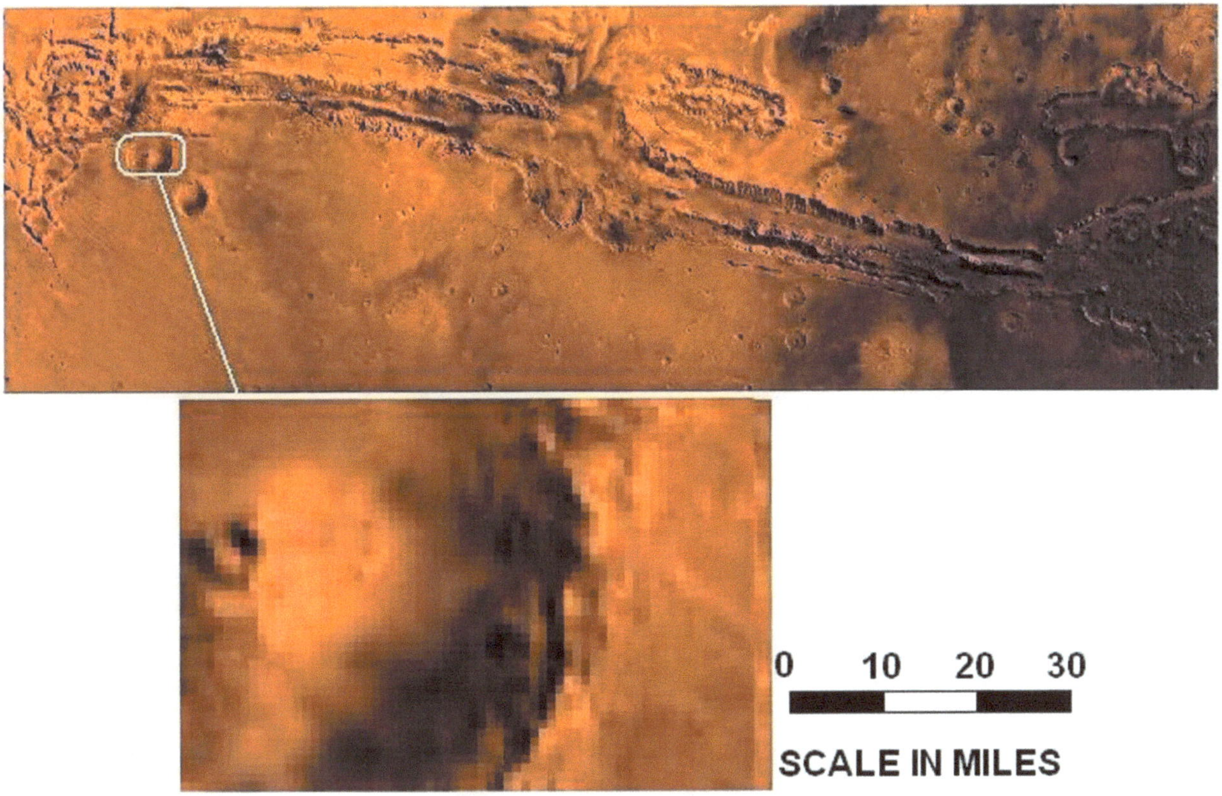

0   10   20   30

SCALE IN MILES

PART 1: LICHENS ON MARS

Lichens are a combination of two life forms: a fungus, called a mycobiont, and algae or bacteria that uses photosynthesis, called a photobiont. Lichens are a unique form of life, not plant, not animal. They do not need water to survive, and they live in some of the most extreme environments on Earth. In 2005 it was learned by experimentation in space that lichens can survive in vacuum, in outer space. On April 26, 2012, an experiment concluded that lichens could survive under laboratory conditions similar to Mars' atmosphere. Therefore, it is not surprising that there are lichens on Mars, regardless that NASA never mentioned them. This photo from the 1997 Pathfinder rover shows that lichens can get quite large on Mars.

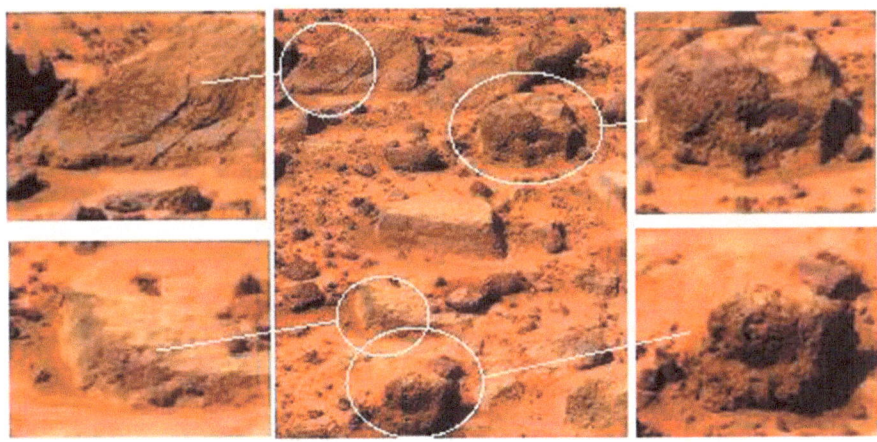

Also, from the Viking Lander 1 in 1976, this green rock is actually covered with lichens.

The Viking Lander 1 wheeled robotic rover, the Sojourner, examined a rock with a lichen on it, and was photographed by the lander, but even after analyzing the rock, the lichen went unnoticed and unmentioned by NASA."

lichen

rover

Just for comparison sake, here are photos of a couple of lichens that I found in my neighborhood, on Earth. These are one or two inches in diameter.

LICHENOMETRY

Lichens are used to determine the length of time stones have been unmoved or not submerged in water, with a method called lichenometry. They seem to grow on stones, masonry, and trees, only after decades. Rocks that have been stationary for about forty years or more may have small lichens on them. Lichens on Earth grow in radius at a slow and constant rate of approximately 25 millimeters in 40 years, some types faster, some slower. At that rate, a lichen like the largest one in the Pathfinder photo, roughly 1500 mm (about five or six feet) in diameter, would be growing for about 1200 years, on Earth. On Mars, they would probably grow more slowly due to the lower amount of oxygen, lower sunlight intensity, and lower temperatures. So the large one in the Pathfinder image may be thousands of years old.

PART 2: ANIMALS ON MARS

"Life on Mars" presented a number of plants and animals from Mars photos from NASA's Phoenix, Spirit, and Curiosity rovers. The plants included a red-leafed plant that hugs the surface, a pointed leaf plant like a wandering Jew, a mushroom, and flowering plants that have flowers like lilies, pansies, and petunias. Photographs showing approximately two dozen species of animals were also presented. This book adds about a dozen animals to the list of animals photographed on Mars. The photographs of animals included the following:

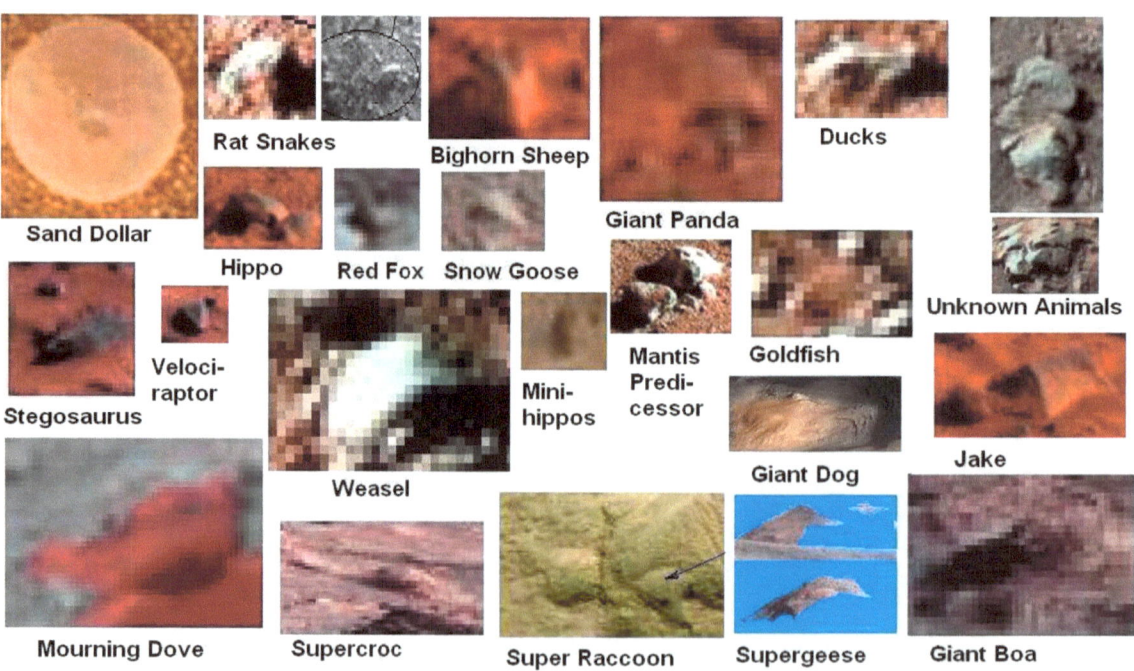

BRONTOSAURUS

These four images from the Pathfinder rover are essentially the same image taken seconds apart and from different cameras. They show what I believe is a brontosaurus, which was a dinosaur that lived on earth from 154 to 150 million years ago...about the same time as the stegosaurus. They look similar, except for the plates on the back of the stegosaurus. The long tail seems to be hidden in the sand.

 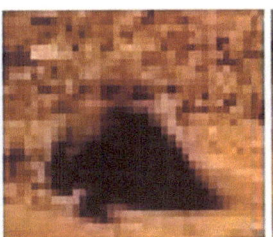  

TURTOCEROUS

These animals look like some kind of turtle-rinocerous mixture, so I call them turtocerouses. They are probably male and female. One type has a concave back, one has a convex back. There are several in the photo and in similar photos taken by the Pathfinder rover at that site.

KANGAROO

This is the head of a big kangaroo sticking out of the sand, with eyes closed. This was one of the rare cases that NASA took more than one photo of the same thing, so that movement could be assessed. In this case, you can see that the head of the kangaroo has moved. Also, notice that the head of the turtocerous has changed position from one photo to the next.

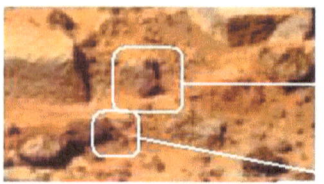   

BROKEN-ROCK BUNNIES

These two things look like broken rocks but they have legs, heads, and bodies something like huge big-footed rabbits. I call them broken-rock bunnies.

ROCK MOUTHS

The photo below looks like huge rocklike mouths that come up out of the desert, as if they wait for things to come into their mouths and then they close their mouths. One of them has some kind of animal or bird on it's nose.

BLUE MOUSE

Is this a mouse (or similar rodent) that knows how to camoflauge itself like lapis lazuli? Or is it a piece of lapis that is shaped to look like a mouse? Or is it just the way of NASA photos? A similar mouse appeared in a Viking Lander photo, again, looking like a stone, (far right).

13

ROCKNEST MONSTER

From the Rocknest photo that shows the search effort for someone lost in the sand, this big black thing looks like a rock but then again it has eyes, a nose, and mouth, and legs. It is an animal that looks kind of like an elephant. Notice how it looks at the camera while a human searcher is coming out of the sand behind him. Also other humans (searchers) are behind him, and although it is hard to see and hard to believe, it appears that there are kids on his back and his feet. Interesting babysitter. This animal seems to be highly intelligent, harmless, and easily domesticated, in spite of it's gruesome, monstrous appearance. Its huge feet seem to have evolved to allow it to walk on the loose quicksand without sinking into it.

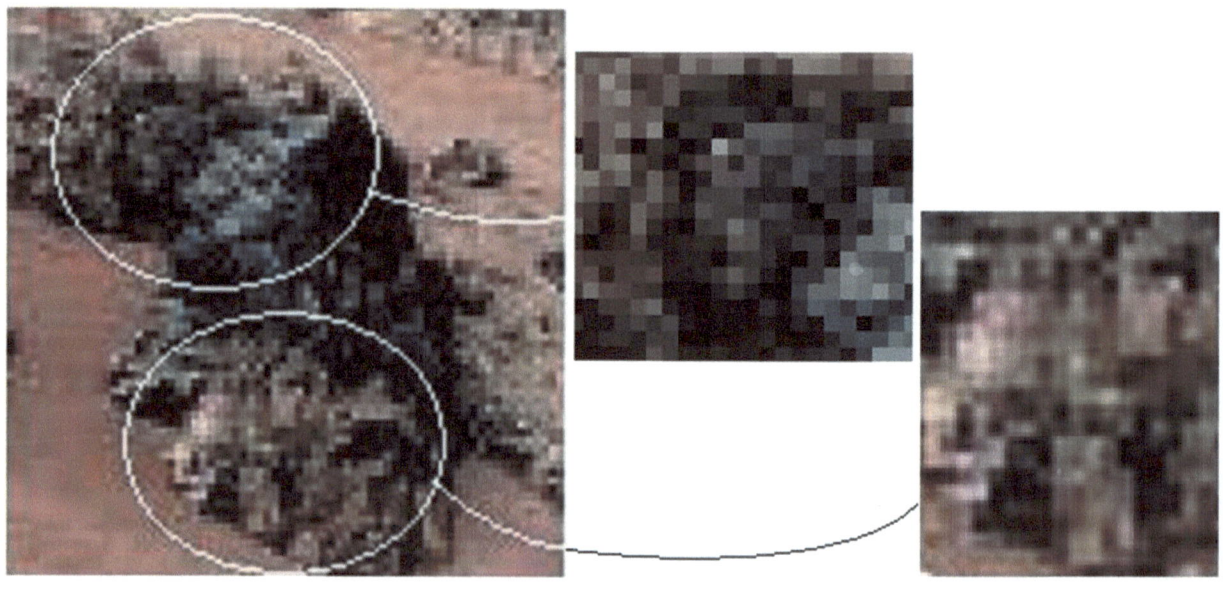

MORE ABOUT THE FOX PUP ON METEORITE

If the fox pup head, (shown initially in "Life on Mars"), is four inches long, then the meteorite is 9.5 feet across. NASA claims that the meteorite is basketball size. A standard basketball is 9.5 inches diameter. This says something about the accuracy of NASA's scale. Additionally, there is something in the sand in the distance beyond the meteorite, which I think may be a doberman pinzer.

PTERODACTYL

Curiosity rover photographed what NASA calls the edge of Yellowknife Bay, on 12/27/2012. The photo shows a vast sandy area between the rover and the ultimate objective, Mount Sharp, several miles distant. In the sand these three images of animal heads can be seen.

One of the creatures, on the far left, looks like the head of a pterodactyl, which was a flying reptile that existed on Earth, 150.8-148.5 million years ago, as determined by fossils. That is the same time that Brontosaurus and Stegosaurus existed on Earth. The other heads are unidentifiable.

 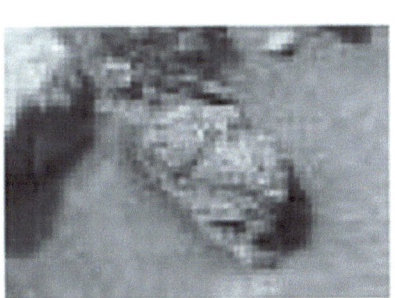 

SUPER-DUPER CROC

In the satellite image of Gale Crater, which shows a giant dog, there is also a huge supercroc in the lake. The lower legs and tail are visible, but the body and head are submerged in the lake. The extra-large supercroc is about 75 miles long, and probably consumes any animals or fish that venture into the lake.

PART 3: MARTIANS

"Life on Mars" presented a weak case for humans on Mars. The following images were presented:

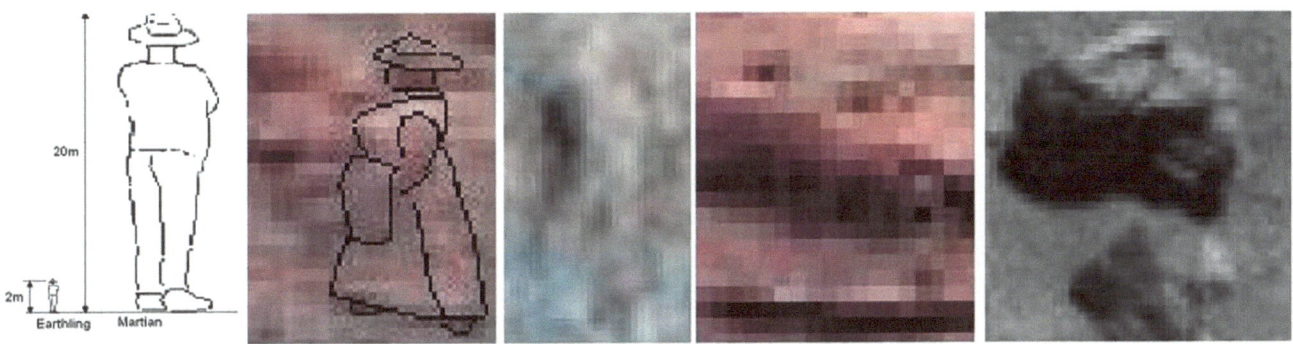

However, due to their blurry and ghostlike composition, and poor resolution, they fall just a hair short of convincingly illustrating that Martian people do in fact exist. They are useful, but not totally obvious to the degree necessary. They show that there are recreational activities, (for example, the supercroc farm), but also that there are hazards, (dry quicksand pits, animals). Some also show the Martians as gigantic size, consistently, 20 meters tall.

So far, not one individual has told me that they see and believe the images of water, plants, and animals, in my first book, "Life on Mars". People look at the photos and seem to be unable to focus their eyes on them. Although the images are just too unclear for conclusion by most people, I find them clear and obvious. I find that discrepancy frustrating, but I am convinced totally that Martians do exist, and therefore, I feel compelled to document and share these discoveries.

Who, or what, are the Martians? They are not little green men. They are not reptilians. They are not giant insects with super intelligence. Astoundingly, they are humans, who probably were swept from Earth to Mars in tremendous floods as the two planets collided or brushed,thousands of years ago. They are like us, with bodies and minds, just like ours. At this time, I can tell you, from the images available, that there are three kinds...which I would call "normals", "giants"and "supergiants". All are shaped proportional to us on Earth, and look and dress like us, except for their enormous size. The normals are the same size as we are. It seems that all kinds wear clothing which is either modern or traditional, which I would refer to as "modern" and "ancient", but also there are at least two races, black and caucasian. They have cities and may number in the millions.The supergiants are apparently only rare, with perhaps hundreds or so on the planet, approximately 100 to 300 miles tall, too tall to compare in the sketch below; even a toe might be a mile across.

The size of the giants was determined by scale references by NASA on the images that they were seen. There is a possibility that the scale is erroneous, but I have no way to check it or verify it. The twenty meter height is approximate, but it is from at least three independent image sources, so it seems to be genuine.

Giants   Normals

The Martians have the same, or very similar, ability to reason that we do. They know how to build complex structures like buildings, roads, dams, and bridges. They know metalwork. They have cars and roadbuilding equipment. They have compressed air cylinders. They have rubber or plastic tubing or hose. It appears that they do not have our modern ways like computers, phones, or cameras, but they do seem to have electrical devices. Their level of technology seems to be limited, something like our 19th Century, but with some 20th Century developments. Their energy source is unknown, although I suspect one possiblity is methane biofuel. There are no signs of pollution like from fossil fuels or even trees, on Earth. I have not seen fire but I have seen forged metal. There are no signs of electric wires or wireless transmitters. There are no planes, at least in the air. They seem to be quite simple compared to ourselves. They appear to be so much like us, it seems as if they came from Earth. Perhaps they may know more about Earth than we know about Mars, (judging from their modern clothing, backpacks, and hats), but at this time there is no evidence of astronomy tools such as telescopes. They may have come from Earth thousands of years ago, and so there could be legends about Earth. The ancient normals may be descended from ancient Egyptians, as they wear clothing similar to them. If they originated on Earth they may think if it as "home".

THREE MARTIAN SIZES

There are three general Martian sizes, although there may be others, or ranges. This may have been caused by different intensities of magnetic field zones, or it may be that some people simply did not protect themselves as well from the cosmic solar radiation. Perhaps some of the first humans to have been transferred to Mars did not realize that the intense radiation caused giganticism, and, because they lived on the surface, they evolved over many generations to become larger and larger with each succeeding generation. The ones who burrowed and lived under the ground then did not grow to gigantic size.

NORMALS

Most of the people on Mars are normal sized. The best examples of a normal-sized human from the rover photos are:

GIANTS

As I explained in "Life on Mars", the humans that were seen in images in Gale Crater were 20 meters tall, using the scaling information that NASA provided. Some of those humans were dressed like modern Earth humans, but also some of them wore ancient Egyptian clothing.

SUPERGIANTS

There are humans on Mars that have grown to incredible sizes. Similar to the hundred-mile dog in Gale Crater, there are people that gigantic, as well. The bald-headed, elderly man with black clothing and spectacles measures ninety miles across at the shoulders! The man crawling in the crater with blue jeans scales sixty miles tall. One has to wonder, how does he get his clothing? The little girl's head in the crater photo at the end of the Introduction scales thirty miles wide, which would mean she is roughly 300 miles tall. How tall will she be when she grows up? And how does she dig a hole to stand in?

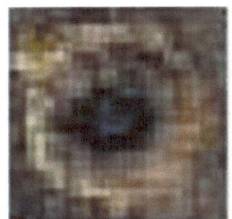  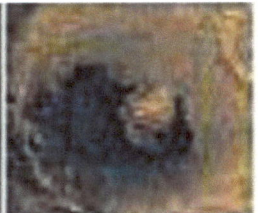 

THE CITY

The following enlargement from a Curiosity photo shows the city. You can see the road in the lower portion of the enlargement, with the bridge visible at the middle of the photo. The bridge appears to be ramped upwards, from right to left, but this is probably just an illusion due to the perspective angle of view. The waterway must be level. There are many buildings and structures in the photo, demonstrating that Martians have developed capabilities very similar to our own modern construction capabilities.

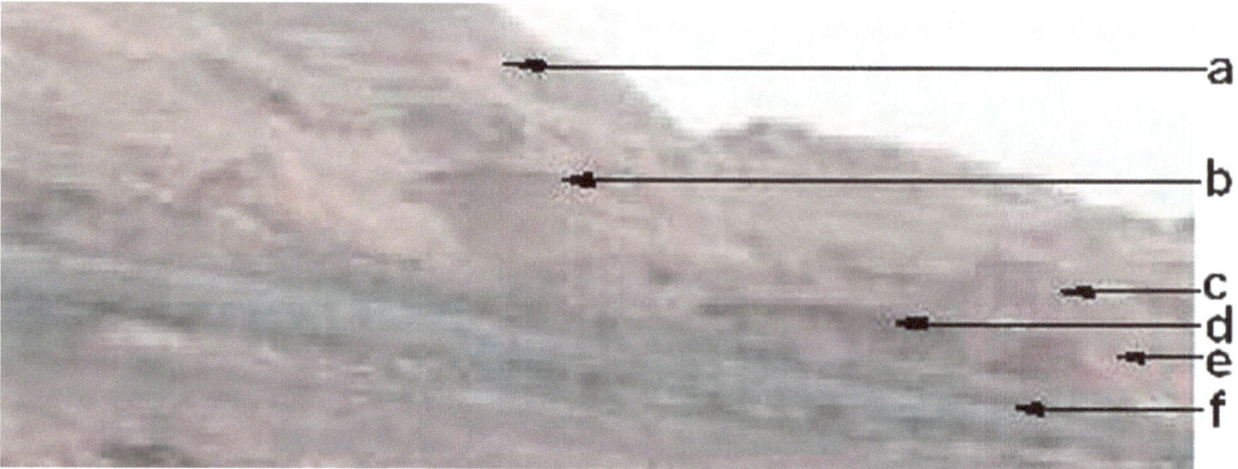

"a" in the photo appears to be numerous buildings on or near the top of the cliffs. The edge of the Gale Crater forms the cliffs. These buildings follow the contour of the cliff with several terraced levels, and have window strips, just like modern Earth office buildings; they likely manufacture and use glass to appreciate the view and allow sunlight. For office buildings in modern Earth cities, the floor-to-floor height is roughly double the average height of humans, and therefore I believe they would have this same feature. Space is needed for air conditioning ductwork, structure, and clearance for various human activities. These facilities appear to be for normal-sized humans, as the large amphitheater (d) is about 300 meters diameter.

"b" in the photo is a circular structure, with a diameter roughly four times the floor-to-floor height of the building in "a"; therefore, it is around 150 meters diameter, not much different from a basketball arena or sports stadium on Earth. It might be a stadium. A stadium that size might be for normals, but not for giants.

"c" appears to be a rectangular structure, several stories in height, sitting on a solid rectangular base. In this building, the windows are continuous floor-to-floor, like curtainwall construction, so the floor levels are not visible. The buildings appear to be concrete construction.

"d" is another large circular structure, possibly another stadium, although much larger than "b", roughly twice the size, or 300 meters diameter. This might be some kind of arena, like the ancients had for chariot races.

"e" is a smaller building, one of many in the image.

"f" is the road. There are no cars or vehicles on the road. However, there are some faint, blurry ghostlike images which might be giants, walking along the road.

There may be many cities or towns all around the perimeter cliffs of Gale Crater, and elsewhere on the planet. This can only be verified with future photographs. The general problems with the rover program seem to be coming to awareness and attention now, that, other than the lack of clarity of the photos, the rovers in the past have landed in the midst of relatively low populated areas, in remote wilderness. This is probably what would happen if rovers landed haphazardly, by chance, on Earth. Only after traveling across the surface of Mars for many miles, does it become clear that there are societies with buildings and activities of the Martian people. The satellite photos also seem to not be able to show clearly enough what is going on, due to the long distances involved in the imagery. It is quite intriguing, however, that NASA's investigations have been unable to see entire cities.

Imagine a scientific probe from another planet landing in the Mojave Desert, ten miles from Las Vegas, leading alien scientists to conclude that Earth was lifeless. Doesn't that seem absurd?

moose   horse   dog   unknown animal

dog   cow   woman   boy   old man   man   bird

woman's hat   boy's crown   man's crown

ROCKNEST

There looks like a sort of dark brown animal in the background. In the front of the moose you can see some kind of animal with black stripes on it's side. There is a person (a woman) with a hat like a sombrero with a halo sitting next to it on the ground. Maybe that person is trying to milk the animal? And perhaps the animal is a cow!

If you look in front of that person you will see a small person like a child standing there also with something like a crown on his head. You can see him only from the waist up because there is another person in some kind of chair or blanket-covered rocks in front of him. The person in the chair seems to be looking to his left and it appears to be an old man, with a white cap, and kind of skinny. He may be sitting on an elaborate throne. And then another man, clothed only in short pants or a wrap, and a similar white cap, looks like he is trying to move some big thing, possibly some kind of giant unknown species of animal. There are some other animal forms in the photo, including dogs, as well as a horse, standing behind the old man.

These people are seemingly nomadic travelers, living in the open, using the desert rocks like furniture, apparently without any kind of permanent residence or even portable dwelling. Or possibly, they are involved in the search for a person lost in the sand, but they live somewhere else. I suspect that they sleep with thick blankets. They are not heavily clothed, as if the temperatures are mild. It is spring on Mars now.

The old man in this photo seems to be lying on a heavy blanket of some sort, probably wool, and also there is a striped blanket on the back of the cow that the woman is milking. Another person may be sleeping under the same heavy blanket, at the man's feet. This is probably dawn, as the sun is low. And still another man, lightly clothed, muscular, is pushing a giant snake or something like a snake, thirty feet to the right.

There are probably similar humans on Earth in this sort of condition, nomadic persons who roam the desert, wearing little clothing, for example, in remote villages of third world countries. And similarly, on Earth, there are modern people who wear clothing like those at the dam and at Glenelg. So the situation on Mars is quite similar to that on Earth, sociologically. There are city folk and country folk.

The part of this that is extremely confusing is that according to scientists, the temperature gets down to something like minus 140 degrees F at night, and rises to 80 F only at mid-day in the summer. The people do not appear to be in an environment that cold. A human on Earth would freeze under those conditions, and they would wear thick, insulating clothing including gloves and hats. In Antarctica, for example, people survive only with due precautions against the cold. So something seems awry about this entire scene. Either the temperatures are mild as they appear, or the people are different from us.

And on top of that mystery of the light clothing in frigid temperatures, there is the question about the air they breathe. 96% Carbon dioxide, .14% oxygen. That in itself is a very strange mystery. The only explanation I can think of is that humans have the ability to acclimate to the temperatures and low oxygen levels, and animals also seem to have that ability.

The hat of the woman looks like a sombrero with some kind of circular "halo" on top. The halo might mean something about her status. She is milking a cow so I think she may be a woman servant or daughter to the old man, who I suspect is royalty of some sort. She also has a necklace, and wears a wrap that covers her chest and waist, down to the knees. The young boy seems to have something on his head as well, circular in shape, which I suppose is a crown, which could mean he is a prince or royalty. The man pushing the snake has a white cap, a much less pronounced head covering, and he is wearing nothing but a sort of wrap around his waist, similar to those worn by the ancient Egyptians, called shendyts. The old man lying on the blanket appears to have some kind of elaborate throne, a spectacular, large ornamental thing, stretching several feet above his head, glistening with rainbows, in the sunlight. Maybe it is decorated with polished metals and jewels. Perhaps he is some kind of royalty. And then the man trying to move the snake seems like a man-servant. He is very muscular and has no fear of the giant animal he is trying to coax away from the camp. The snakelike animal could be the animal depicted on the ancient Egyptian Narmer Tablet, a sort of lionlike creature with head like a snake, called a serpopard, (serpent-leopard).

car

swimmer

animal

dog

canoes

swimmers

king

car

searchers

guards

THE ROCKNEST SEARCH SITE

At the search site, Rocknest, the photo shows that there were divers apparently looking for someone lost in the quicksand. They used canoes to float about in the sand, and they seem to have used some kind of diving equipment including air tanks and inflated bags, hoses, and diving suits with caps, nose guards, and air filters. They have something like plastic tubing for the air supply hoses.

Also at the site, there is an ancient normal man who looks like a priest or royal person with a crown, sitting down, with a heavy blanket wrapped around him, facing some kind of altar, and a possible throne in the area behind him. Nearby, a female modern normal diver watches two male modern normal divers in a pool of quicksand. In the distance, people swim in the sand, leaving wakes similar to wakes in water, while dogs also join in the search. There are abandoned cars and discarded tires in the area. My suspicion is that the missing person is related to the royalty. Here is a sketch showing the man with a crown and blanket, facing an altar.

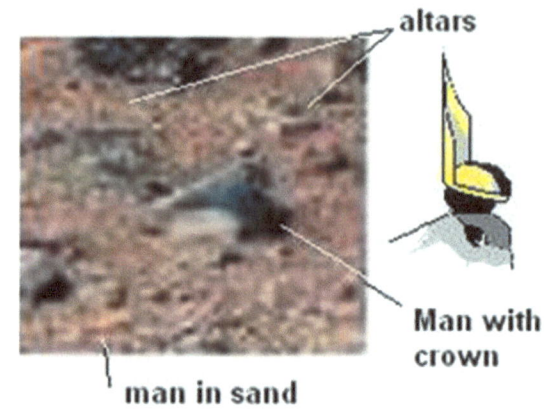

The altars suggest that this person is a high priest. Also, it suggests religious ritual.

The ancient Egyptian kings wore a similar crown, as depicted on the Narmer Palette. Interestingly, the palette shows animals called sherpopards, which are legendary combination snake/leopard. The king is also shown as a giant on that palette. Could it be that the Egyptian pharoah depicted on the tablet was a giant human from Mars?

ROCKNEST SEARCH DIVER

In a very strange quirk of fate, the rover photographed the head of a diver in the sand on October 15, 2012, when it scooped a sand sample. Then again, it scooped and photographed the same head in the sand, (he has to be dead), on November 9, 2012, 25 days later, from the exact same position. The two pictures can be compared, as seen below:

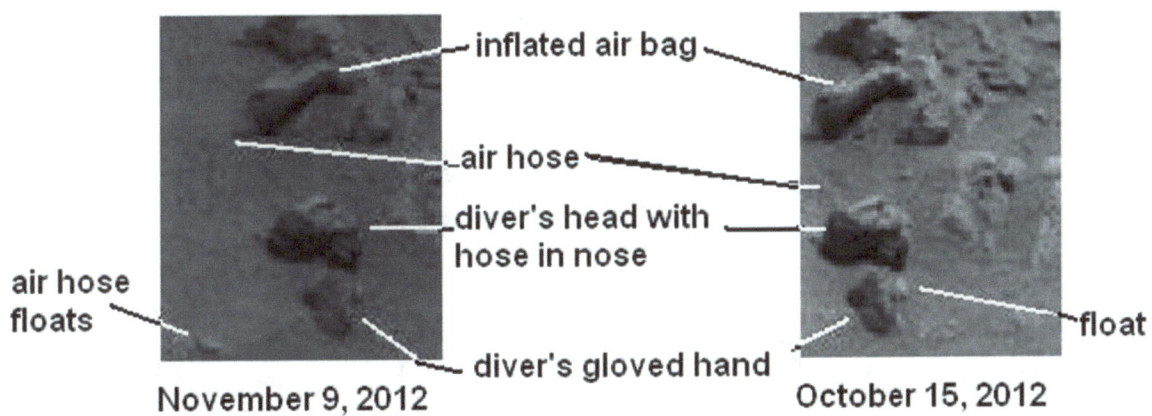

November 9, 2012   October 15, 2012

I can imagine that a skeptic might say that this is not a human head and glove in the sand...it is just a rock. After all, it has not even moved or changed negligibly in 25 days. But I would ask, then, why is it that the picture shows an inflated bag, with an air hose, and some kind of white floats, leading from the air bag to the nostrils of the diver? What kind of a rock is that?

That brings up an interesting question. What would happen if, (and when, since it is quite likely), Martians discover the rover? My guess is that they would disable it. What would we do with an object like that if we found it in the desert? We would probably have our military take it to a lab and take it apart. Don't you think? As I wrote in "Life on Mars", the moment is near that we will interact

with Martians. Yet we have essentially no way to communicate with them.

MARTIAN TECHNOLOGY

Some of the rover photographs give clues to the level of technological development of the Martian people. A good example is this picture of what looks like a front-end loader, road building equipment. The type of fuel used is not known, but the stantion next to it could be an electric charging station. As can be seen, the equipment is black in color.

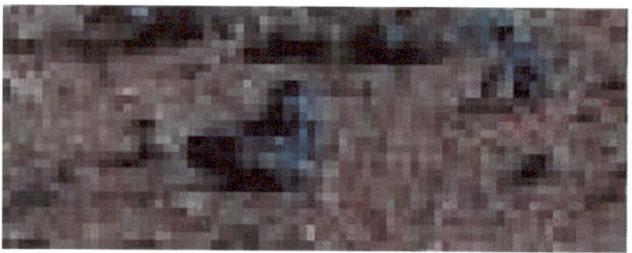 

Also, the photo from a distance of the Rocknest area, by Curiosity rover, shows people diving with dive suits with breathing apparatus

priest or pharoah with blanket in front of altar

female search team diver in sand

diver in sand compressed air cylirder

Another possible car photo may have been the Curiosity photo which showed a church like object.

possible church

possible car

Cars seem to exist but in very limited numbers. The photo showing an elephant "parked" next to a car suggests that people use a variety of animals for transportation by some, and cars by some elite few.

car

car
horse

elephant

car

From the 1997 Pathfinder photo, there is a metallic structure, possibly a vehicle, and also possibly an entrance to an underground space. The object is white metal, like aluminum. There is a person who is sitting in the end of it, apparently opening or closing some semicircular doors. There is a window

also. The person has long braided ponytail and a bra, shirtless. There is a tailgate also. My guess is that this is some kind of safety station that the woman uses to protect herself from the environment, but also, it could be a subterranean entry point. This object, whatever it is, again demonstrates, (like the cylindrical metal objects in the Viking Lander photos, and the cylindrical tank and pipes in the Curiosity photos), that they have a sort of cylindrical mindset in terms of design and construction. Oddly, however, in the midst of all the cylindrically-designed objects, is the 1949 Chevy Fleetwood at Rocknest. Someone on Mars collects cars?

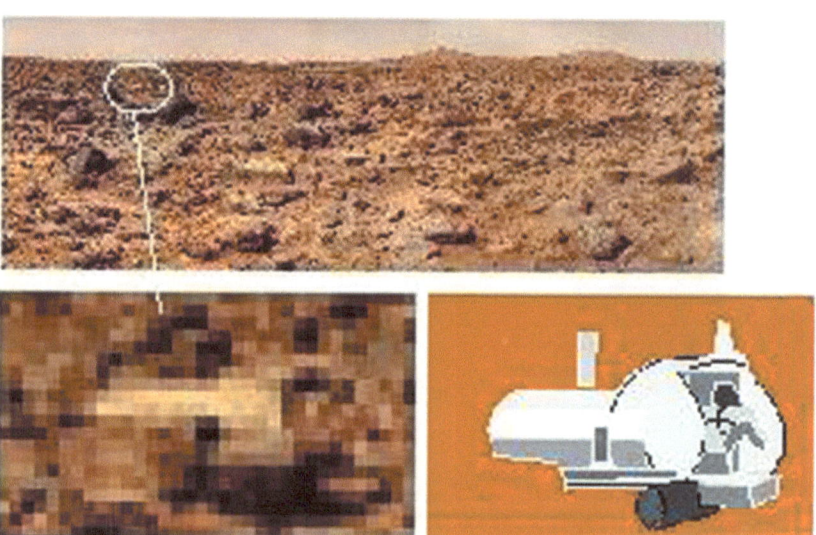

ANCIENT EGYPTIAN CULTURE

Several things point to ancient Egyptian culture. First, there is the crown on the man sitting before an altar, which is very similar to those worn by pharoahs in ancient Egypt, for example, the Narmer

Pallete. Secondly, there is the cloth kilt worn by the man pushing a snakelike creature, which is similar to those shown in ancient Egyptian hyeroglyphics, called a shenydt. Third, there is the helmet, Viking Lander image, which is similar to ancient Egyptian helmets, depicted in Egyptian hyeroglyphics. Fourth, there is an altar image, which looks like a bull deity again relates to the Narmer Pallete bull dieties, from approximately 3000 BC.

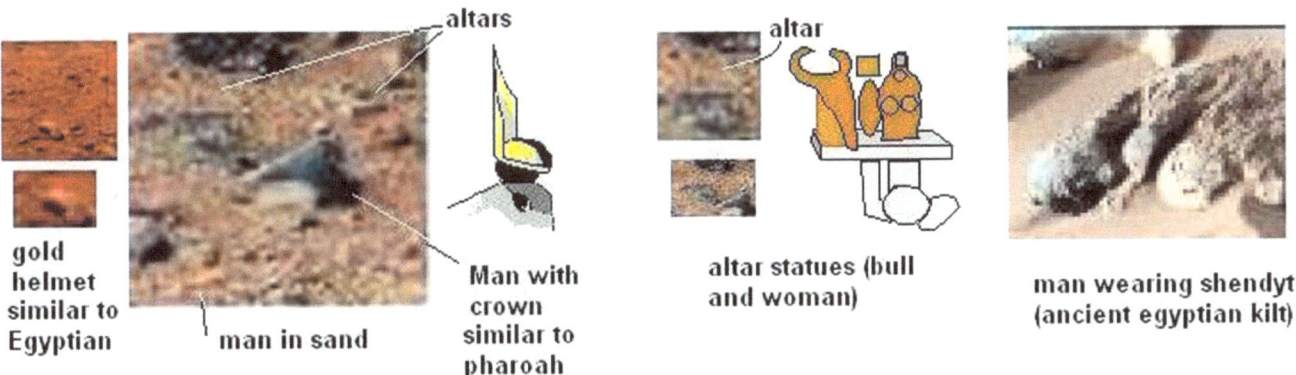

gold helmet similar to Egyptian

man in sand

Man with crown similar to pharoah

altars

altar

altar statues (bull and woman)

man wearing shendyt (ancient egyptian kilt)

MYTHOLOGICAL GIANTS

There are stories of giant humans in ancient times, all over the Earth. Could it be that they were actually more than just myths and legends and fictional stories? Zeus, Hercules, the Cyclops, and other Greek deities and their offspring, Ara, the soldier whose body (by legend) was cast on Mount Ararat, Goliath, slain by David, the giant humans that built Nan Madol in the Pacific, perhaps the Olmecs whose giant head statues are in Mexico, the Easter Island statues, and other stories of giant humans, all beg to be restudied in the light of the giants on Mars.

PART 4: EARLY MARS EXPLORATION

BRIEF HISTORY OF MARS EXPLORATION

Prior to the invention of the telescope, astronomers/astrologers depended entirely on eyesight for planetary observations. Mars was seen and tracked. It's motion was observed. It's red color tone was noticed. It was considered a malevolent planet, affecting humans in negative ways; for example it was the Roman god of war. Even today modern astrologers, who base their understanding of planets on ancient knowledge, consider it to be an influence that leads to egotistical interactions, like competition.

Glass lenses for reading glasses were used in the fifteenth century, paving the way for the invention of the telescope in the early seventeenth century. Although Galileo is thought of as the inventor, he simply improved it and applied it to great extent, focusing primarily on Jupiter and Saturn.

As early as 1651, Italian astronomer/theologian Giovanni Battista used a telescope to define all the planets known at that time, including Mars.

In 1877, another Italian astronomer, Giovanni Schiaparelli, used an 8.6 inch refractor telescope to make a map of features on Mars. He named many of the features that are observed even today. He published a map of Mars, showing channels, which were later deemed to be optical illusions, and claiming that it was inhabitable.

Percival Lovell built an observatory in Arizona and studied Mars, writing three books (1895-1906) about the surface features of Mars, depicting canals and suggesting intelligent life on the planet, similar to Schiaparelli's conclusions.

In 1908, Orson Wells wrote "War of the Worlds" which was a fictional account of an invasion by Martians, based on Schiaparelli's work. Our knowledge of Mars remained imaginary, vague and controversial for over half a century, afterwards.

The first successful modern mission to Mars was not a rover but a flyby, named Mariner 4. Mariner 2 went to Venus successfully. Mariner 4 took pictures from a high altitude of the surface of Mars on July 14 and 15, 1965. This was the first advance since telescope observations. Mariners 6, 7, and 9, in the sixties, also were successful in taking flyby photos of Mars and relaying them to Earth.

Many early attempts to land a rover to take photographs on Mars, in the sixties and early seventies, by both the Soviet Union, and the United States, failed. The USSR program started in 1966, but after a series of failures, they gave up. The first successful landing and transmittal of pictures from the surface from rovers were from NASA's Viking Landers.

Viking 1 landed on July 20, 1976, transmitted until November 11, 1980;

Viking 2 landed on September 5, 1976 and transmitted until April 11, 1980.

The Viking program discovered geological features formed by large amounts of water, in Mars' southern hemisphere. Ancient rivers and stream beds, asteroid craters that seemed formed in mud, and such features suggested that large amounts of water once existed on Mars. The amount was estimated to be ten thousand times as much as the Mississippi River.

The next successful attempt was not until two decades later, the Pathfinder 1997, followed by the Spirit and Opportunity rovers, in 2004, the Phoenix Lander in 2008, and Curiosity, in 2012. Future missions are planned, as well.

Some scientists believe that the Viking tests for life on Mars were successful, including Gilbert V. Levin, who designed the Viking life detection labeled release experiment, and Barry DiGregorio, who researched geomicrobiology. The subject is developing and evolving, but there are problems of lack of disclosure, egotistical fear, and credibility.

In 2008, Andrew D. Basiago, a Vancouver lawyer, publlished a paper on the internet called "The Discovery of Life on Mars" in which he examined NASA photographs (similar to this book), and he also concluded that life exists on Mars, including humanoids. He idealistically believes that the planet Mars is not ours to invade and destroy ecologically, but rather, belongs to the inhabitants, and we must obey unwritten cosmic laws in our impositions. His group, Mars Anomaly Research Society, (MARS) is dedicated to research, disclosure, and education about life on Mars. I agree with some of his observations and conclusions, but not totally. My opinion is that he misjudged and misunderstood much of what he saw, and overlooked many things. He probably did not recognize that the statues and sarcophogi and bodies he saw were from Earth. He mistakenly identified blurred images as strange mutants; he did not see giganticism. For example, in this photo he missed the giant frog's head.

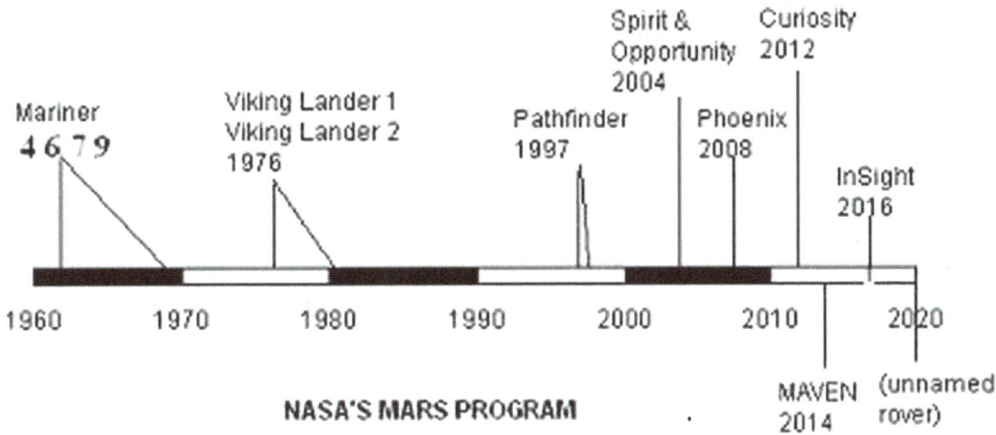

NASA'S MARS PROGRAM

Since the later images, Pathfinder and afterwards, seemed to defy detection of life forms by NASA, even though it was there, there is reason to question whether the early images from the Viking Landers in the '70's likewise defied detection of indications of life on Mars. It seems prudent to at least review some of the images to see what was photographed and if things were misinterpreted. If they do show evidence of life forms, it could be truly revealing and surprising. The first images from those two rovers came back in late 1976.

At that time, the rovers also analyzed soil and showed some indication of chemical evidence of life, but the evidence was questioned as being contamination from the rovers themselves. So, unfortunately, the answer to the question of life on Mars was ultimately deemed to be "no", and that answer has prevailed to present day. For some strange reason, the truth has eluded scientists for decades.

MARINER 4, 1965

The Mariner 4 flyby mission reached Mars on July 14, 1965. It took 21 pictures during the flyby, from an altitude which at lowest point was about 10,000 km, (6,000 miles).

This was the first look at the surface of Mars from a point closer than a telescope on Earth. The appearance disappointed scientists, who saw moonlike craters and barren surface. Subsequent to obtaining the images, most scientists judged that there was little hope for life on Mars other than small, microscopic life forms. That judgement has remained, unchanged, for 57 years, accepted in all practicality, as fact. Even science fiction writers turned to other star systems for their fictional aliens after that photo became public. Mars appeared lifeless, but everyone was wrong...except Schiaparelli.

In preparing the adjoining sketch, I used the following photos, mosaic of frames 11 and 12:

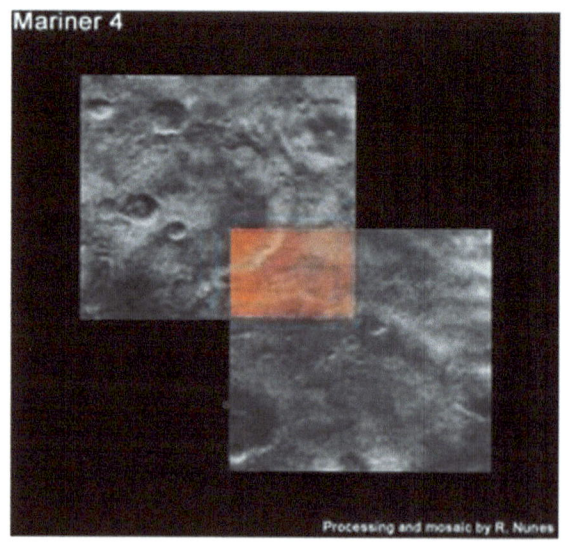

I've outlined what I can see in the adjoining overlay sketch. I've found that by sketching an overlay I can more carefully assess what is in the photo. I started with what looked like an arm, which at first sight also looks like the edge of a crater. By carefully outlining the forms, I eventually discovered the face of the woman on the left. A close and thorough inspection of the photo shows a man lying on top of a woman and also a second woman, next to them. (We've invaded someone's bedroom on Mars!) There are various manmade implements in the photo as well, (spear, gun, bracelets, beads). Is it real or imaginary? My experience with images from Mars tells me it's real.

The temperature at the surface was estimated by Mariner 4 at minus 100 degrees Celsius which converts to -148 degrees Fahrenheit. The angle of view in the photo was 1.05 degrees; therefore the size of the image can be calculated, as the distance from the surface was known to be approximately 6,000 miles. A 1.05 degree angle projected for that distance would show an area approximately 110 miles square: (1/360 X 6000 X 2 X 3.14 X 1.05). That means the human forms in the image were even taller than that, over 150 miles or so! Although that interpretation seems absurd and impossible, assuming that the distance from the surface and angle of view are not erroneous, the human forms would be the size, approximately, of Gale Crater, which roughly scales, according to NASA, the size of Connecticut and Rhode Island combined, which is over 120 miles. So, if you can believe that the dog in Gale Crater, (explained in my first book), is that big, you should be able to also believe that the human forms are that big. That's just another way that Mars stretches our imaginations.

Since the atmosphere on Mars is only ten miles thick, a person that big could not even stand up, for he would enter the vacuum of outer space, and have zero oxygen, which might be fatal. Therefore the giants must keep their bodies and heads low to the surface.

If it is real, then by simple deduction, human forms must be visible on satellite photos of Mars. If we enlarge a map of Mars on our computer screen so that it is roughly 20 cm, representing 4000 miles, then human forms 100 miles tall should be 2 mm tall, roughly, visible on the photo. Furthermore, if

we logically presume that humans that large would need to keep their heads below the atmosphere, and thus, keep either prone to the ground or dig holes to stand, we should look in craters. Here are examples of human forms in craters on Mars, enlarged to assist in judging the images:

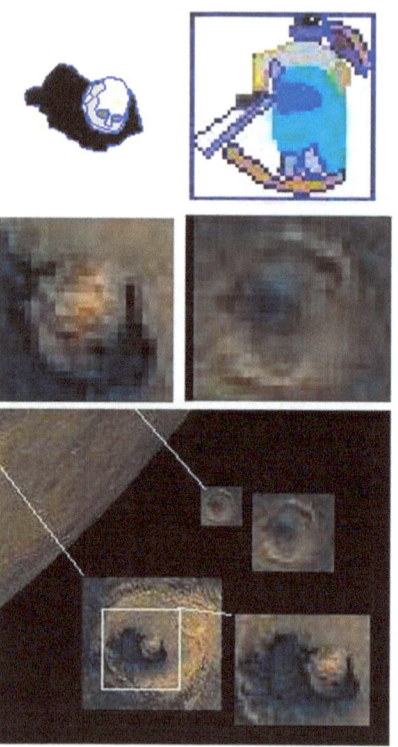

And so, we can see some giant humans on Mars, even without rovers, simply by observing the planet by telescope from flybys and orbiting satellites, simply because they are so big.

The location of the images from Mariner 4 can be determined by the coordinate system on Mars. The following diagram, using NASA map and data, shows how the photos taken in 1965 were very close to the location of Curiosity rover in 2012, assuming that the eleventh picture out of 22 is roughly in the center of the swath of photos.

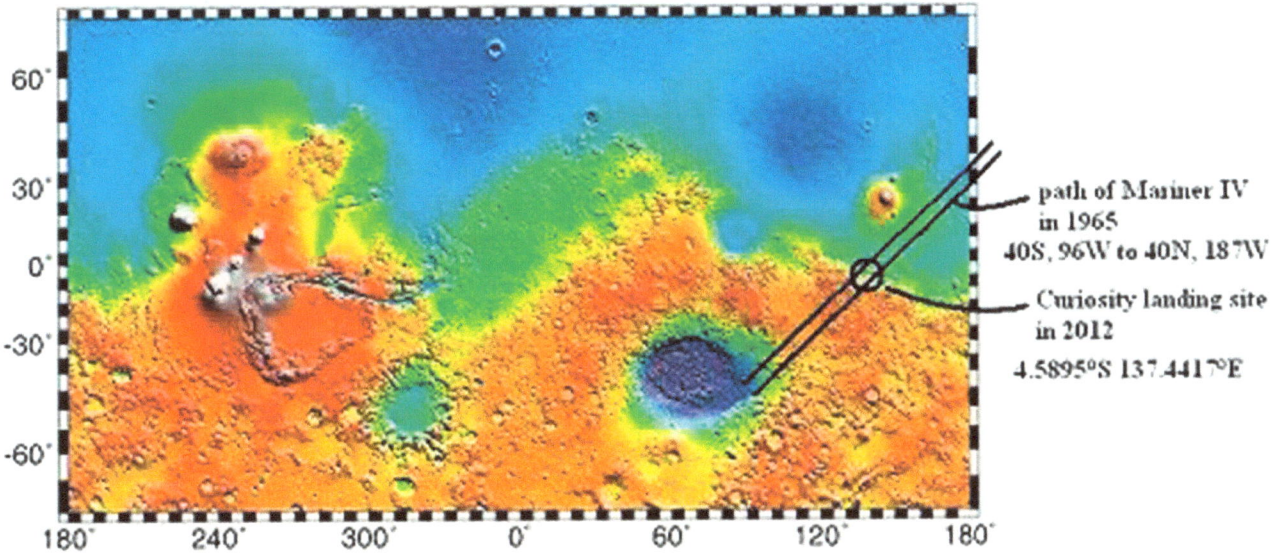

VIKING LANDER 1

The next photo is the very first color photo that was taken by a landed rover on Mars, by Viking Lander 1, soon after landing on July 20, 1976. There are some human forms, possibly a structure, and a few strange objects in the photo

One is a shiny metallic object that looks like a helmet or possibly a pot or dish upside down. It has a yellowish tint to it, like brass or gold. It is flat on top, with sides that come down like ear-protectors on a helmet, similar to ancient Egyptian headwear as depicted in heiroglyphs on the wall of the Temple of Dendera, which dates around 2500 BC. (In reviewing ancient metal helmets, there is a sort of developmental progression, in which helmets after Egyptians became larger and more protective of the nose, face and frontal neck area, unlike the one in the photo.) Also in the photo, at lower right, is what looks like stonework, or statuary. It also looks like a crown of some sort. Still other objects look like manufactured artifacts. To top off the oversights on this photo, it appears that the rover is sitting in a puddle of water. It seems the deserts of Mars are still littered with debris from the great flood, five thousand years ago.

The following was a later image, also taken by Viking Lander 1

There are human forms in this early Viking Lander 1 photo. To be specific, a man has shot another man with a gun while two women remain bound like captives beyond in the sand. One woman is tied to a sort of crossarm like a crucifixion. A line projected from the pistol leads to the approximate location of the head of the dead man if he were still standing. It appears that the dead man was shot just seconds before the photo was taken.

There is also a mouselike form in the same photo, similar to the "blue mouse" inset for comparison; (see PART 2 - ANIMALS). In addition, there are huge, cylindrical shapes in the sand. What are they, explosives bunkers? A military base? Also, there appears to be a huge eye with eyelid, looking out of the sand.

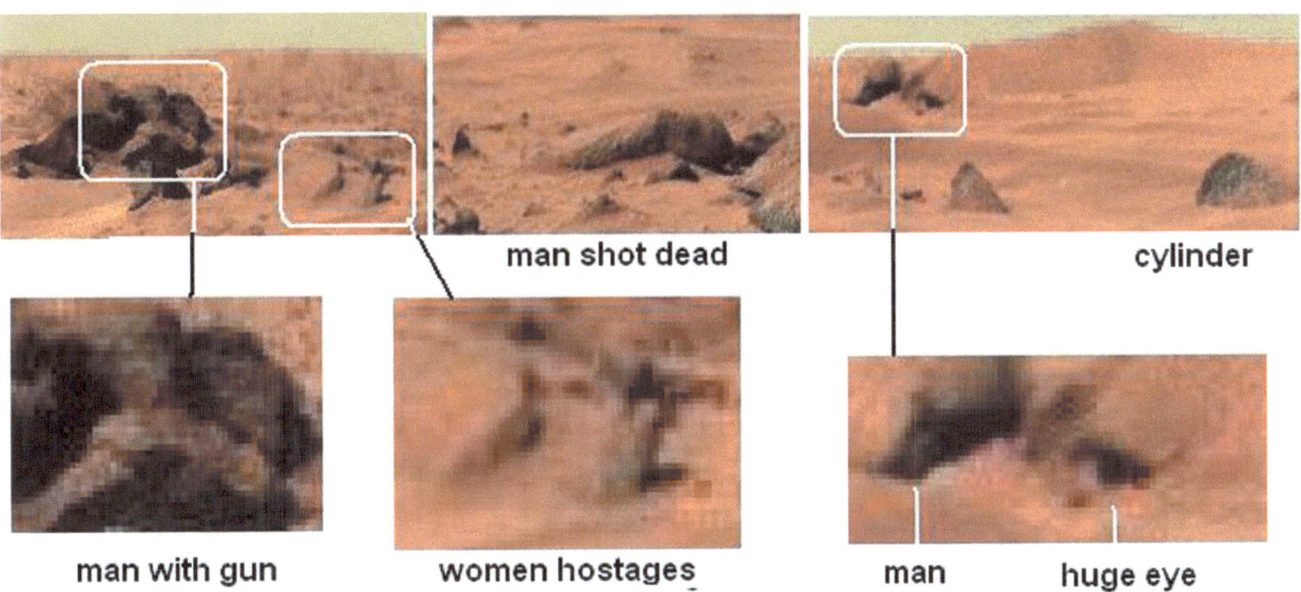

man shot dead

cylinder

man with gun

women hostages

man

huge eye

43

Still another photo from the Viking Lander 1 shows an arched structure like a Mosque, an object something like a modern jet fighter plane, and a Jake-like animal in the foreground. The jet fighter plane has a front intake like a Russian MIG, but an additional intake in the tail. Since there is no runway, this either crash-landed in a field, or is capable of vertical takeoff like a Harrier. It could be simply an object for entertainment, such as in a child's playground. The pilot's canopy appears to be cylindrical. This possibly indicates that jet engines and planes similar to those on Earth can and do exist on Mars." However, their numbers seem limited, since none have been observed (yet) in the sky.

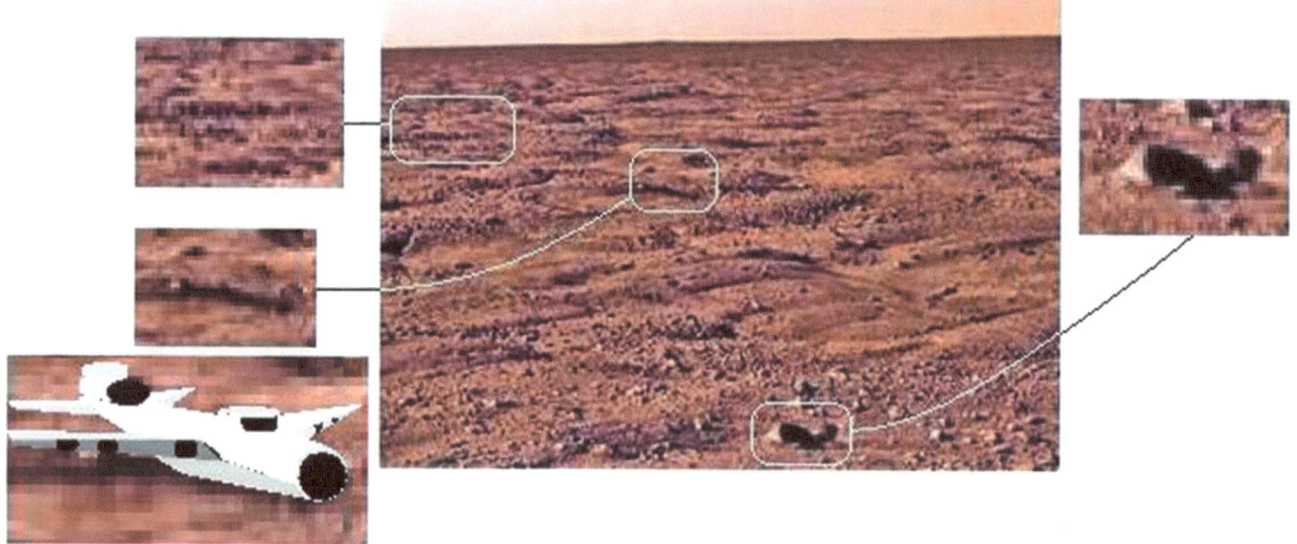

VIKING LANDER 2

This was the first color picture transmitted from Viking Lander 2:

In the above photo, the first image from Viking 2, there are humanlike forms. There appears to be a woman and two children. The woman is seated, facing away from the rover, about forty feet from the rover, fully dressed with a sort of laced cape on her back, and a round cap or wrap on her head A cactus plant flowers at her back with huge round flowers.

In the same photo, but a different, modified lighting version, below, shows a man with a megaphone and speaker, seated on the ground. Many animals surround him, while small pig listens attentively in front of him. Other vague, humanlike forms seem to be present, although barely detectable.

REVIEW OF VIKING LANDER PHOTOS

After reviewing the photos from the Viking Landers, we can say that they had images of a rodent that camoflages itself to look like a rock, a Jake rock, which is an unknown species of animal that looks like a pyramid-shaped rock, several human-manufactured devices, and several human forms.

Looking back 44 years ago, to 1976, it seems possible to me that a close scrutiny of the rover photos at that time would have given credence to the possibility of humans as well as animal life on Mars. It makes one wonder, "Why didn't we see that back then"? Similarly one tends to wonder if the Martians saw the rover. It appears that the rover was within sight of the human forms, within forty feet or so. But it seems as if the rover was invisible. One would think that a moving, metallic object on wheels would attract attention of humans, but it did not. Either that was just a matter of luck or possibly, the humans just accepted it as an unknown stranger in their midst. Either way, the humans on Earth and the humans on Mars continued on, for decades, unaware of each other. Perhaps humans are simply, by nature, oblivious to their surroundings, poor of eyesight, and subject to error.

PART 5: EARTH/MARS EXCHANGES

AGE OF AQUARIUS

The description of the Solar System is quite complex and growing, but it is quite well defined on the internet. As they say, the more you know, the more you know you don't know. That applies very well to astronomy. Nevertheless, I would recommend that anyone with an internet connection, especially children, but really, anyone, get into astronomy by reading internet wikipedias and such sites. Whereas, a few hundred years ago. our knowledge of the solar system was essentially absent, today is different, and anyone who can read can learn and understand it. Additionally, many television documentaries have been made to summarize our latest knowledge of this topic. This is particularly true in the past ten years, so from a historical basis, today's humans can understand the Universe much, much more deeply than anyone has in the past. We are truly in the Age of Aquarius, a period of learning and understanding far superior to any other age in human history.

ORBITS OF EARTH AND MARS

Earth orbits the Sun every year, while Mars orbits the Sun approximately every two Earth-years. Earth is roughly 150 million km from the Sun, which is 1 AU . Mars is about 240 million km from the Sun which is approximately 1.5 AU. The orbits are nearly circular, not exactly; Earth's eccentricity is .017, while Mars' is .093, so Mars' orbit is more elliptical than Earth's. Their path around the Sun is nearly in the same plane; inclination of the Earth's is 7.2 degrees from Sun's equator, while Mars' is 5.6.

As Velikovsky suggests in his book, "Worlds In Collision", the orbit of Mars may have temporarily become elliptical during recent historic times, due to some outside gravitational force, such as a

comet, causing it to veer closely to Earth. We should examine that idea a bit further in detail.

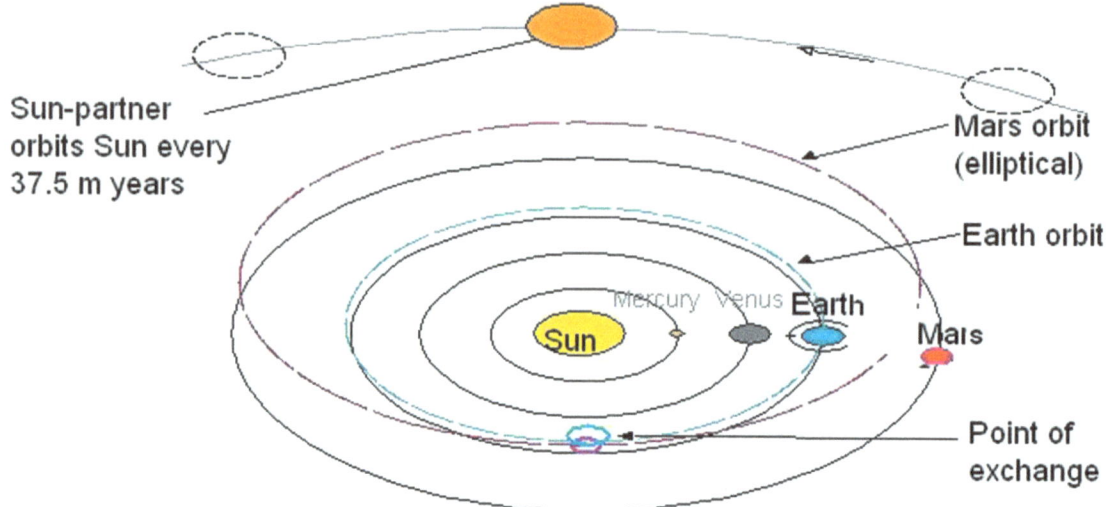

Sun-partner orbits Sun every 37.5 m years

Mars orbit (elliptical)

Earth orbit

Mercury Venus **Earth**

Sun

Mars

Point of exchange

ORBITAL SPEEDS OF PLANETS

If we treat the orbits as circles, we can calculate the velocity of the planets as they travel around the Sun. (Or, we can search the internet.) The velocity of Earth in it's orbit, that is, the speed of Earth as it travels around the Sun, is roughly 66,000 miles per hour (30 km/second). The velocity of Mars in it's orbit is roughly 46,000 miles per hour (24 km/second). Jupiter, by comparison, has an orbital speed of 13 km/second. Saturn's is 9.6, Uranus' is 6.8, Neptune's is 5.3. Pluto's is 4.7. An Oort cloud object called Sedna, discovered in 2003, taking 11,000 years to orbit the Sun, has an orbital speed of 1 km per second. So you can see, by comparison, that as planets get farther from the Sun, their orbital speed declines. Therefore, one can theorize that an object with a great average distance from the Sun would likely have a very small orbital speed. If that object takes millions of years to orbit the

Sun, it would probably move very slowly, less than one km/second, possibly even much less, like a fraction of a km/second, even a matter of meters per second. Humans on Earth can run at a speed of meters per second, so astronomically speaking, that is, to put it mildly, slow, in relation to the speed of the Earth. My point is that if there were an object that orbited the Sun in millions of years, it would take a long time to pass close to the Sun, during it's innermost travel...perhaps centuries...due to it's very slow orbital speed. Therefore it might influence the planets' orbits for many cycles.

DID THE EARTH AND MARS COLLIDE?

If so, then there would be evidence of the collision in the geographical features of both planets. Consider the shapes of the scratch on Mars and the Mediterranean Sea. The two features are about the same size as well as shape. They could be impact "scrapes" in which the two planets brushed against each other, leaving scars or depressions. There are hundreds of cities that have been located by divers under the surface of the Mediterranean. Likewise, on Mars, there are fields still strewn with what look like boats, sarcophogi, bodies, statues, and other archaeological remnants of a gigantic flood. Also, Mount Ararat is the location of Noah's Ark, found in 2008 at 13,000 feet elevation. And the Rock of Gibraltar might have caused a deflection at the end of the scrape on Mars.

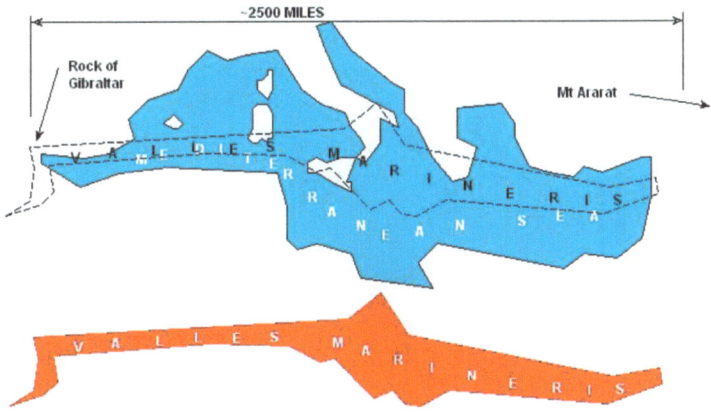

THEORY OF EARTH/MARS EXCHANGES

1) The two planets, Earth and Mars, came together very closely or collided many times in the past.

2) The collisions occurred approximately 37.5 million years apart, on a regular basis, by a sun-partner.

3) The last collision (or series) occurred within the past 60,000 years, possibly during the time of the great flood.

4) During the collisions, life forms, including humans, from Earth, were transferred to Mars, and vice-versa.

5) During the collisions, water on Earth flooded to Mars.

6) During the collisions, natural deformations occurred on both Planets.

The following evidence supports the above theory:

1) Many of the life forms on Earth are identical to those on Mars.

2) Life forms now extinct on Earth exist today on Mars, and some of those existed on Earth during different time spans.

3) Scientific studies indicate that large amounts of water existed on Mars in the past.

4) Historical records (e.g. Bible) indicate there were giant human forms that fell to Earth in the ancient past, "fallen angels".

5) Some Mars life forms exist with gigantic size.

6) Giant human skeletons have been found on Earth.

7) Historical records indicate great floods and natural deformations on Earth.

8) Shape of scratch on Mars matches shape/size of Mediterranean Sea on Earth.

NEMESIS (HYPOTHETICAL SUN-PARTNER)

In 1984 a theory postulated the existence of a Sun Partner, (most stars have twin partners that orbit each other), a dim star that orbits the Sun with a period of 26 million years. The star has never been seen, but is postulated as a cause of regular extinction events on Earth, eleven periods, during which extinctions seemed to accelerate, roughly 26 million years apart. Hence it is named "Nemesis". Nobody knows exactly how big or even where it is. Even the best telescopes cannot see it, but that does not mean it does not exist, because, in theory, it is not emitting light, is too far away, and simply, has not yet been found, and may never be. There is a dispute amongst astronomers whether it actually exists or not. Maybe it's invisible because it's transparent. After all, it is just a blob of helium and hydrogen, both of which are invisible in gaseous form.

PERIODICITY OF EXCHANGES

The findings of animals on Mars in "Life on Mars" tends to support the theory of Nemesis, except that the theory of periodic extinction does not work. The relevant animals on Mars have been determined to have been living on Earth as follows:

Humans: sixty thousand years to present
Animals on Earth currently, in general (e.g. Panda): Pleistocene and Holocene Eras (2.5 million years ago to present)
Sand Dollars: 50 million years to present
Velociraptor: 75-71 million years ago (extinct)
Super Croc: 112 million years ago (extinct)
Stegosaurus: 155-150 million years ago (extinct)

From this information we can plot on a scale covering the past 200 million years on Earth, as shown:

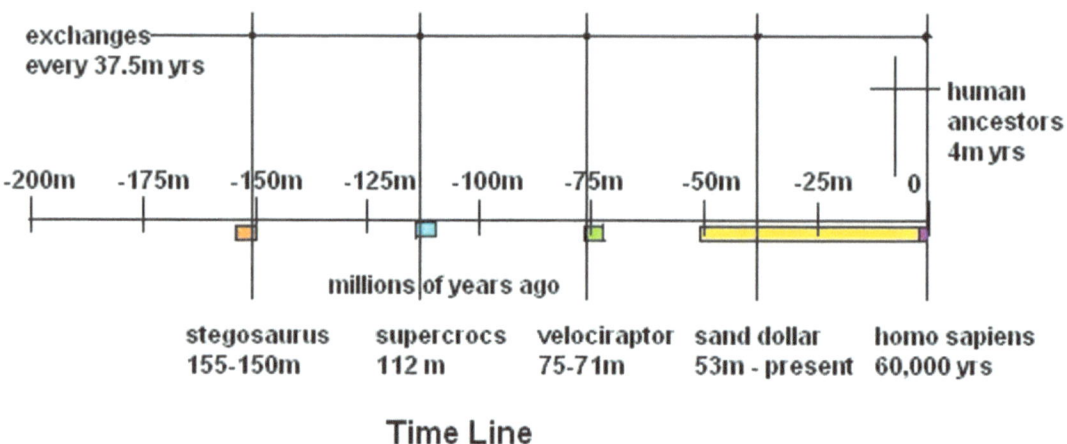

Time Line

If we consider the 26 million year cycle, we see that it does not quite fit. For example, Velociraptor lived from 75-71 million years ago, but if modern humans experienced an exchange within the past 60 thousand years, then the prior Nemesis passes would have been <.06, 26.06, 52.06, and 78.06 million years ago, too late and too early for the velociraptors. But the 37.5 million year cycle fits, since the passes would have been <.06, 37.56, and 75.06 million years ago. Therefore, for the purpose of this book, I will theorize that the Nemesis cycle is 37.5 million years, and the last pass was within sixty thousand years, and, as I will explain, during the time of Noah, within the past five to ten thousand years.

Let's say there was a Nemesis pass through the inner or outer Solar System sometime during the past ten thousand years. That would explain homo-sapiens as well as the bighorn sheep, fox, weasel, dogs, horses, cows, and panda, being on Mars today, since those mammals evolved within the past several million years during the Pliocene era, and were different prior to that. The mourning dove

also probably fits into that category. Strangely, some animals on Mars now exist in isolated areas on Earth, such as the panda, in China, and the mourning dove, in North America. What does it mean? It MIGHT mean that those animals originated on Mars and fell to Earth. It also MIGHT mean that they fell from Earth to Mars. Note that they are in the northern hemisphere.

DID NOAH'S ARK LAND ON MARS?

Noah's Ark was found, with 99.9% certainty, at elevation 13,000 feet on Mount Ararat, in Turkey, in 2010. Radiocarbon dating put it at 2700 BC Therefore, the animals and humans on Mars did not get there by Noah's efforts. More likely, they swam.

According to the Bible, Noah's ark coincided with the fallen angels, godlike fathers of the Nephilim, who were giants compared to Earthlings. Noah gathered up animals as best he could, but of course not every species, only those that were readily accessible to him. So how did he find species in North America or in China? He probably did not. Did Noah travel to North America to gather species? I doubt that.

Did the animals come from Mars to Earth, rather than vice versa? This does seem feasible. For example, the gravity of Earth is greater than the gravity of Mars, and therefore it seems that things from Mars would fall to Earth rather than vice-versa. It is difficult to understand exactly how something could "fall" from Earth to Mars, considering the lesser gravity of Mars. One possibility does come to mind. Suppose the two planets did come together, almost touching or scraping gently. All the water would flow to the point between them, as in this diagram:

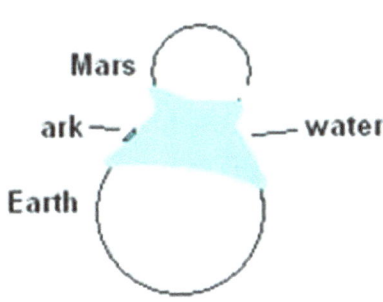

What seems to work, fitting all of the examples, is the 37.5 million year cycle. If we assume a pass within the past 60 thousand years, that is during homo sapien time. At 37.5 million years it was during the sand dollar time. Then 75 million years ago was during the velociraptor, then 112.5 million years ago was during supercroc, and then 150 million years ago was during stegosaurus, brontosaurus, and pterodactyl.

I realize this is scanty evidence, and that it involves very limited data, and much more in depth study is needed. My goal is just to get the idea started, just to lay the general basis for the theory.

THE NEPHILIM

The Bible, in Genesis, states that the Nephilim were sons of gods that fell to Earth. Also, it states that the gods were forbidden to marry humans. So there were other angels, who came to Earth and managed the extermination or deaths of the violators who intermarried. This story suggests, if it has any bearing on the theory of exchanges, that there were at least two times that Mars and Earth came together at near the time of Noah, (First there were fallen angels, then their exterminators.). The idea seems possible if the Sun-Partner, Nemesis, took many years to pass by the Sun and in it's elliptical orbit, caused the interplanetary meeting twice, once on approaching the Sun, and once upon leaving. Also, the Book of Enoch, which was excluded from the Bible, had some information about the "Fallen Angels", their offspring, the Nephilim, and the Watchers and their progeny. Also, there is an excellent modern accounting by Scott Roberts and Craig Hines, titled "The Rise and Fall of the Nephilim" (2012). Hopefully, this discovery of giant humans on Mars helps to understand and believe such tales, myths, and writings.

CONCLUSIONS

My first book, Life on Mars, concluded that there was water, plant life, animal life, and a high likelihood of humans, on Mars. Evidence of humans was presented including a dam, roads, pipes, tanks, and faint human images; however, the evidence regarding humans fell short of conclusion due to doubt about the clarity of the photos.

This second book goes further. It shows that many of the animals that have evolved on Mars, looking much like rocks, probably as a defense mechanism. The rocklike animals have fooled Earthlings for decades, because their camoflage is so well developed. Also, clearer images of humans have eliminated all doubt of their existence.

It also goes further in showing that the humans on Mars, Martians, have some comfort but difficulties within a harsh environment. Rover photographs show that there are cities, social structure, and technological developments on the planet. Many findings indicate that the people have a society similar to ancient Egypt, in addition to a city-based society. Some Martians are gigantic, others are normal sized like ourselves. Some are dark-skinned, with Negroid features, others light-skinned with Caucasian features.

Early photographs of Mars, starting with the Mariner Mission in 1965, as well as rover photos from the Viking Landers in the '70's, have demonstrated that there is life, including humans, on the red planet. Those poor quality early photos from Mars rovers, flybys, and satellites, were either not seen, ignored, or passed off as optical illusions, inconclusive, distorted, not convincible, or unreadable images. Decisions by the scientists in response to the early photos, led to misjudgments that shaped public knowledge for decades. In a classic case of the blind leading the blind, once a judgement was made by the experts in the field, nobody questioned that Mars was lifeless.

The truth is that the photos from all the NASA rovers are a gold mine, a treasure trove, of information about life on Mars. Implications to be drawn from the photos are immense and innumerable. Future Mars missions need to take note of the existence of animals and humans there. We cannot expect to continue to keep sending rovers and making scientific exploration without crossing paths with and interfering with the existing inhabitants, eventually.

We need to embrace our brothers in space. We need to establish communications with them, by sending equipment and aids such as microphones, speakers, and computer monitors. We need to teach them our language and learn theirs. We need to learn from them what they know and how they think about humankind and our common history and ancestry. We also need to explore in depth the plants and animals on Mars. We need to search through satellite photos for the giant humans and other animals not yet discovered by this cursory review. We need to study and learn about them.

We need to explore the myths and legends of giants on Earth with the possibility in our minds that they were Martians. The Nephilim, the Greek gods and their offspring, the various stories around the world such as Ara at Mount Ararat, Goliath, Hercules, the giant humans that built Nan Madol in the Pacific, the giants in German myths, and on and on.

If we are considering, in the future, bringing back soil samples to Earth from Mars, we need to also consider bringing back animal specimens, plant specimens, and possibly even DNA samples of Martians. Imagine the excitement of bringing back a DNA from dinosaur or unknown species. Before sending humans to Mars we need to interact with Martians to ensure that we are accepted and protected by them as much as possible. They can help us and we can help them to understand our identities and history. We need to think of Mars, not as a faraway planet, but as an extension of our own.

We Earthlings need to avoid devastating Mars and the Martians as we have exploited and degraded the planet Earth, as we have devastated and strived to dominate other cultures in the past. Hopefully, we will learn from our mistakes and will avoid repeating our destructive history, because, after all, it IS self-evident that all men are created equal, (regardless of size).

Hopefully, scientists will some day see that Mars is a planet covered with life forms, including a human society with ancient ties to Earth, as well as modern ways. It may take years or decades, but some day, Martians and Earthlings will meet and learn from each other. In the meantime, I will continue to explore and share my findings.

Do you think there could be some kind of microscopic life there?

www.ingramcontent.com/pod-product-compliance
Lightning Source LLC
Chambersburg PA
CBHW050943200526
45172CB00020B/535